LIFE ON THE
ISLANDS

Written by **Rosanne Hooper**

Consultant: Roger Hammond
Director of Living Earth

World Book Ecology

Published in the United States and Canada by
World Book, Inc.
233 N. Michigan Ave.
Suite 2000
Chicago, IL 60601
in association with Two-Can Publishing

© Two-Can Publishing, 2001

**For information on other World Book products, call 1-800-WORLDBK (967-5325),
or visit us at our Web site at http://www.worldbook.com**

ISBN: 0-7166-5227-7 (Life on the Islands)
LC: 2001091765

Printed in China

1 2 3 4 5 6 7 8 9 10 05 04 03 02 01

Photograph Credits:
p.4-5 The Image Bank/Marcel Isy-Schwart p.7 (top) Robert Harding/Financial Times (bottom) Robert Harding/Robert Francis p.8 (top) The Image Bank/Lawrence Fried
(bottom) OSF/W. Gregory Brown, Animals Animals p.9 (bottom right) Science Photo Library/Simon Fraser p.10 Robert Harding/Krafft p.11 Bruce Coleman/Konrad
Wothe p.12 Bruce Coleman/ John Fennell p.13 (top left) Bruce Coleman/Alan Root (right) Bruce Coleman/A.J. Deane p.14 Bruce Coleman/Christian Zuber p.15 Survival
Anglia/Dieter & Mary Plage p.17 (top) The Image Bank/Stockphotos/Rainer Kiedrowski (bottom left) The Image Bank/Steve Satushek (bottom right) Survival Anglia/Alan
Root p.18-19 ARDEA LONDON/Jean-Paul Ferrero p.10 (top right) Robert Harding/Brian Hawkes p.20 The Image Bank/James H. Carmichael, Jr. P.21 The Image
Bank/Lynn M. Stone p.22-23 ARDEA LONDON/Jean-Paul Ferrero p.23 (right) ARDEA LONDON/Ron & Valerie Taylor

Front cover: NHPA/W.S. Paton Back cover: Bruce Coleman/Gerald Cubitt

Illustrations by Michaela Stewart. Story by Claire Watts. Design by Belinda Webster. Edited by Monica Byles.

CONTENTS

All words marked in **bold** can be found in the glossary.

LOOKING AT ISLANDS

Islands are completely surrounded by water. There are small and large islands, many of them alive with strange animals and plants. The isolated conditions on an island lead creatures to **adapt** in different ways.

Many develop into new **species**. Every island has many different kinds of **terrain**. For example, an island can have **tropical** and dry forests, desert, mountains, rivers and lakes.

Even bare, windswept islands that look empty have many hidden inhabitants. But because islands are small, there are few members of each species, and so they are more easily in danger of **extinction**.

▼ Tahiti is ringed by a **lagoon** and **coral** reef. If the island sinks, the reef will stay as an **atoll**.

MYSTERY ISLAND

Old legends tell of a lost island called Atlantis. An entire civilization is said to have lived there. Scholars now believe it may have been an Aegean island, destroyed by a volcano over 3,000 years ago.

WHERE IN THE WORLD?

Islands are scattered all over the world. Most are concentrated around the world's coastlines. Up to 30,000 islands lie in the Pacific Ocean alone. Lining the edge of the Pacific is a necklace of volcanic islands called the ring of fire. Among the Hawaiian islands is a chain of 107 volcanoes, over 2,175 miles (3,500 km) long. Another long arc stretches from Alaska across to Asia.

Thousands of islands are clustered in the northern part of the world, some of them without names. A few islands, like Aldabra in the Indian Ocean, are far from any **continent**, and too small to be seen on most maps.

Islands on main routes between continents have become trading or military bases, and **migration** stopovers for birds. Large nations often try to win possession of islands, because of their useful locations and **natural resources**. The Virgin Islands, on the edge of the Atlantic Ocean, officially belong to France, the United States and Britain. But many islands are claiming their independence from the distant countries that own them.

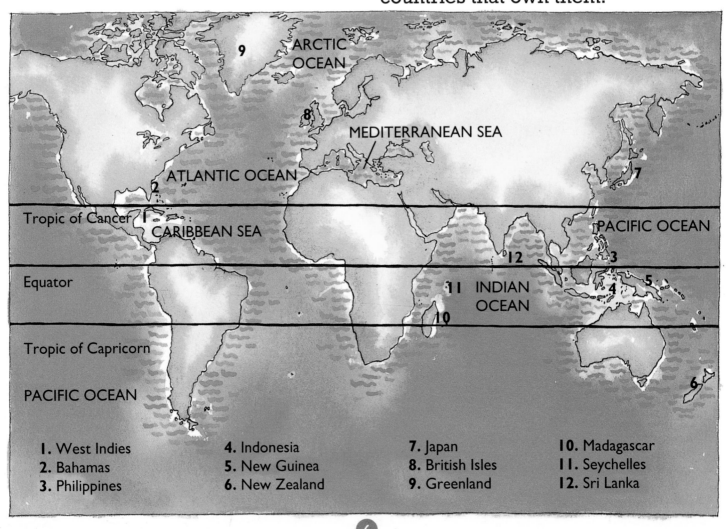

1. West Indies
2. Bahamas
3. Philippines
4. Indonesia
5. New Guinea
6. New Zealand
7. Japan
8. British Isles
9. Greenland
10. Madagascar
11. Seychelles
12. Sri Lanka

▼ Islands come in all shapes and sizes. Some, like these strange Chocolate Hills in the Philippines, were formed when volcanic **lava** bubbled up from under the seas.

▲ Some of the major cities in the world are built in islands. Much of New York City is crammed on to Manhattan Island and Long Island. Skyscrapers are built to save space.

THE BIRTH OF ISLANDS

Islands come and go. Over time, old ones disappear and new ones form. Most large islands are continental: they were once attached to the mainland, but were cut off millions of years ago when the sea level rose after an **Ice Age**. The British Isles were a part of Europe. Islands like New Zealand and Greenland split off from their nearest landmass when the world's continents drifted apart.

Some islands explode into life. Some 20 years ago, a fierce volcano erupted off Iceland. Surtsey Island emerged over a few days from the boiling ocean. Volcanic islands often pop up one after the other and form a chain. Coral islands can form in shallow water around volcanoes.

▲ Smoke billows up from the crater of a volcanic island off Iceland. Volcanoes on Iceland itself still erupt centuries after its formation.

▼ The wooded humps of the volcanic Palau Islands rise sharply from the Pacific Ocean. People have lived on these islands for centuries, growing tropical vegetables and fruit. A coral reef surrounds the islands.

HOW AN ISLAND IS MADE

Volcanic islands
When a volcano erupts on the ocean floor, lava piles up. As it rises above sea level, a red-hot island forms. Nothing can live there until the land cools down.

Coral islands
Sometimes coral forms a ring around a volcanic island. As the volcano sinks, the coral rises, and waves dump sand in the middle, forming a coral island.

Continental islands
When the **glaciers** melted after the Ice Age, the sea flooded low coastlines. High patches of land stood out above the waves and remained as islands.

▶ Geysers are often found on volcanic islands. They produce tall jets of steam and boiling water, and can be used to generate electricity.

GREENING OF ISLANDS

Brand-new islands are barren, with no vegetation. Slowly seeds arrive, blown by the wind or floating in on the tide. Many are carried in birds' feathers, beaks or droppings.

The more remote islands receive fewer seeds, but from one species, hundreds of new ones may **evolve**. Hawaii is a mass of exotic flowers, even though the original seeds had to travel nearly 1,864 miles (3,000 km).

Island plants evolve in strange ways. Plants that are small shrubs on the mainland may grow into huge trees on islands. Plants may even be different on next-door islands. In the Bahamas, one island has orchids not found on any of the other islands.

▲ Silver sword plants on Hawaii are making the most of island life. On the mainland they grow only in fertile forests. These island species flourish even on dry, volcanic soil.

MONSTER PLANTS

Small plants on the mainland may turn into giants on an island where there is less competition for light and water. Sunflowers on the Galapagos Islands in the Pacific have grown to enormous heights.

The world's biggest seeds are the football-sized coconuts from the coco-de-mer tree. They grow only in the Seychelles because they are too heavy to float away anywhere else.

▼ Baobab trees soak up water during the rains and store it in their spongy trunks for when it is dry. Madagascar, off Africa, has nine species of baobab. Just one species grows in Africa.

BIRD LANDS

Islands are natural nesting sites for seabirds that spend the rest of the year feeding out at sea. Penguins and puffins swarm onto islands in the Atlantic. Tropical islands dotting the Pacific and Indian Oceans hum with a variety of birds.

Birds that live on islands all year round develop special adaptations – for example, different beak shapes to deal with the food on the island. Fourteen kinds of finch are unique to the Galapagos Islands. Mainland finches eat only seeds, but island finches include nut eaters with parrot-like beaks, insect eaters with long, thin beaks, and fruit eaters with fat beaks.

It is hard for birds to reach islands from the mainland. They are blown by the wind, travel by boat or arrive on driftwood. For this reason, there are few species of land birds living on the more remote islands.

▼ The pukeko lives in New Zealand. Because once there were no **predators** to escape from and plenty of food to eat on the ground, the bird lost its ability to fly. Today, pet cats and dogs have hunted it close to extinction.

▲ This woodpecking finch on the Galapagos Islands has developed a way of digging grubs out of holes in trees. Its tongue is too short to reach the grubs, so it uses a long thorn.

▶ In Indonesia's Irian Jaya, birds of paradise spread their tail feathers to attract a mate. They even hang upside down from the branches!

DID YOU KNOW?

● The male blue-footed booby on the Galapagos Islands makes a fake nest to attract a female. She lays her eggs in the dust. The pair share jobs: she catches fish, he feeds them to the chicks.

DWARFS AND GIANTS

Island creatures have adapted to suit their **habitats**. On Komodo Island in Indonesia, there are giant lizards called Komodo dragons that eat pigs, deer and goats. In Hawaii, butterflies are huge and heavy so that they are not blown away by the wind. Island creatures can grow big because there is little competition for food and no need to hide from predators. There are also tiny animals, such as the little sika deer of Japan, which survive because they need so little food.

It is easier for insects, bats and birds to reach islands than it is for animals with no wings. Once creatures do arrive, they often change form. In Hawaii, up to 200 species of tree snails have evolved from just one mainland variety.

CHRISTMAS CRABS

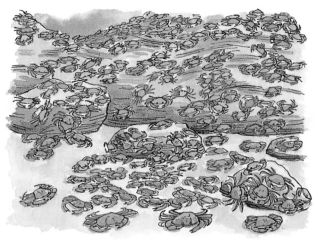

Each year in the Christmas Islands off Indonesia, small red crabs march to the coast. After mating, the males turn back inland. The females remain, until one day at high tide, they scramble down the rocks and shake their eggs into the water. Each crab lays about one million eggs. If only two stay alive, the species survives.

◀ Giant tortoises on the Galapagos Islands are the largest in the world. They grow to over three feet (one metre) long and can be as heavy as four people. Sometimes their heads grow so big they cannot hide in their shells.

▶ Marine iguanas look a lot like dinosaurs. They move very slowly and spend up to an hour at a time munching seaweed underwater.

ISLANDERS

Native island people live closely with the land and sea. They take only what they need to survive and make good use of their resources. On the Gilbert Islands in the Pacific, islanders use coconuts for food and water, for building materials, and even to make ropes for fishing.

Each island has its own fishing techniques. On Sri Lanka, fishermen sit all day on stilts in the water with just a rod and line. Sometimes the fishing is a chance for a social get-together. On Fiji, the people from two villages may meet to put out their nets together. In Samoa, locals all gather on the beach at the November full moon. As night falls, they catch fat palolo worms breeding in the sea and feast on them.

PEOPLE PROFILE

Carnival comes to Trinidad once a year. People dress up, dance and make drums out of oil barrels. They sing *calypsos*, funny stories that they make up as they go along.

Long ago, the Viking people lived on the Shetland Islands, in Scotland. Their descendants still build a Viking longship every winter. The ship is paraded, then burned.

◀ Ancient people made their own islands to live on in Lake Titicaca, Peru. They heaped soil onto large anchored reed mats. Local people still build their houses and boats from reeds. They add new reeds as the old ones rot.

▼ The Asmat people of Irian Jaya are fierce warriors. This man is wearing a painted clay mask to perform an ancient dance. Sometimes the Asmat wear bones of their ancestors to give them extra strength.

▼ The markets in Indonesia sell fruits and vegetables. Local people use about 4,000 plant species for food, medicine, and shelter.

INVASIONS

Plants and animals are at risk from the strong winds that blow across an island, but the worst enemies are human. Early settlers brought in foreign animals that ate away whole habitats. Land iguanas on Santa Fe in the Galapagos Islands were only saved from extinction by killing the imported goats. Dogs and cats eat birds, while rats eat tortoises and snails. In Hawaii, rare plants are threatened by beetles from China.

Farming changes the landscape. On Mauritius in the Indian Ocean, forests were cleared for sugarcane **plantations**. Many native species disappeared. Vacation resorts also disturb wildlife, destroy habitats and **pollute** the area. Toxic waste from industry kills off life on and around island shores.

Easter Island, off Chile, was once covered with lush forests. Thousands of years ago, the trees were cut down by the people who lived there. The soil was left dry and barren and the people were forced to find new land. Today, only these ancient statues remain.

STRIPPED BARE

Pacific islands are in danger. Landing strips for military aircraft have squeezed out wildlife from some small islands. After foreign countries tested nuclear bombs, 14 islands were left uninhabited and six were completely destroyed. Other islands are losing their forests fast.

ON THE BRINK

The rare creatures found on islands are irreplaceable. When species die out they do not come back. Some species are already extinct, while hundreds are on the brink.

Of all the birds that have become extinct in the last 200 years, most lived on islands. White starlings live on Bali in Indonesia. Only 250 remain in the wild, but they are still being hunted.

Many animals disappear as their forest homes are cut down. Even local people are not safe. Today, the Punan tribe of Kalimantan in Indonesia fight for survival. If action is taken now, some of those **endangered** may survive.

▲ Tropical birdwing butterflies are running out of flowers. Farmers are destroying their homes to create commercial palm-oil plantations.

▶ Orangutans live deep in the lush rain forests on the islands of Borneo and Sumatra. People are cutting down much of the forests, and so many orangutans are being taken to live in nature preserves to stop them becoming extinct.

ON THE ENDANGERED LIST

The majestic Philippine monkey-eating eagle is one of the world's largest birds of prey. People hunt them, sell them to zoos, and cut down their forest. Only 500 are left in the world.

The huge kakapo parrot cannot fly, because until recently there were no dangerous predators in New Zealand. Pet cats have now eaten so many kakapo that only 60 remain.

SAVING THE ISLANDS

Conservationists and wildlife groups worldwide are working to protect endangered island species of both plants and animals. A wide variety of **environmental** organizations campaign to reduce pollution, military activity, and the destruction of island forests.

Many governments have banned trading of species at risk, such as parrots, turtles, and creatures that live in seashells. International laws also protect some areas of special importance, such as the Galapagos Islands.

Some groups have had success in breeding endangered animals. At Jersey Zoo, some Mauritius pink pigeon chicks have been hatched. There were no more than 15 of these birds left in the wild.

On the island of St. Helena in the Atlantic Ocean, schools, scientists,

WHAT YOU CAN DO

● Join a conservation organization or local group. You could become a **volunteer** and work directly to help the endangered plants and animals of the world's islands.

● Encourage your whole family to buy and use environment-friendly products at home. This will help to reduce the pollution that spills into the sea from waste pipes and factories.

● Tell your friends what is happening so that they can take action, too.

● Leave nesting birds alone on coastal walks, and take good care not to damage plants or animals. Report any sick or injured animal you find to an adult.

▶ This New Zealand national park is one of the few rain forests in the world that is unchanged. The area is protected from any interference by humans.

conservationists, the government, and the local inhabitants have all joined forces to try to save their native plant life.

Animals and plants on islands all over the world will remain in danger until people stop clearing their forests and dumping waste.

▶ This coral island in the tropics is a protected marine preserve. Tourists are only allowed to visit certain areas. The rest of the island is left undisturbed so that the fish can live in peace and the coral is not damaged.

THE MAGIC FISH HOOK

For thousands of years, people have told stories about the world around them. Often, these stories try to explain something that people do not really understand, such as how the world began. This story, told by the Maori people of New Zealand, is about the creation of their island.

Maui had been adopted as a baby by the god of the sea and brought up under the waves. When he was grown up, he left the sea and went back to the land that he had come from. He was always boasting to his brothers and his wife that, because of his upbringing, he was half a god. But Maui was very lazy. While his four brothers went out in their fishing boats every day to catch food for their families, Maui would go out only once a week.

The rest of the time, he lay around at home, getting under his wife's feet.

"Oh, Maui," she would sigh, after she had tripped over him for the tenth time, "Why are you so lazy? Why can't you act more like your four brothers?"

What an unreasonable thing to say! Maui might lie around most of the time, but, on the days that he did go out fishing, he had extraordinary luck, for the fish just seemed to jump into his boat. He caught more fish in one day than his brothers caught in a week. So, whenever his impatient wife began to complain, Maui would just nod and busily start to mend his nets. If his wife happened to come back a few minutes later, she would find Maui in his favorite spot again, dozing in the warm sunshine.

One day, however, one of the four brothers overheard Maui's wife complaining.

"Call yourself half a god..." she sneered. "Well, you're certainly only half a man."

The brother ran off and told the others, who thought it was a great joke. They all began to tease Maui about being only half a man. Maui was absolutely furious.

"I'll show them," he stormed, and went off to a secret spot in the hills. There he made a bone fishhook, carving it with the skill he had learned under the sea. When it was ready, he polished the fishhook and whispered magic words over it.

The next morning, as his brothers were getting ready to go fishing, Maui came up to them. "I think I'll come with you today," he said casually. So they all set out to sea in their boats. A little way out from the shore, the brothers stopped rowing. Maui looked surprised.

"The warm shallows here are full of fish," they explained.

"Who needs fish? I want to go much further out and catch something much better than fish!" Maui boasted.

Maui's brothers were curious to find out what he was talking about, so they all paddled far out into the sparkling ocean.

"Stop," said Maui, putting down his paddle. "I think it's about here."

With that, he dangled his fishhook over the edge of his boat. He felt around for a few minutes and then pulled. The line went taut. Maui's four brothers gasped.

Maui pulled and pulled at the line. What on earth could Maui catch out here? As he tugged, the ocean began to roll and swell, as if some enormous thing was slowly rising up from the depths of the seabed.

Maui's brothers grabbed the sides of their boats to stop from falling out. Then all of a sudden, they and their boats were resting on dry land.

Maui had dragged a whole island from the bottom of the ocean.

"I knew it was around here somewhere," said Maui, climbing out of his boat. "I must just go and apologize to the god of the sea for taking some of his land." And he set off, striding across the grass.

Maui's brothers sat in their boats with their mouths open. Then they began to get out and cautiously put their feet down on the turf.

"I declare myself king of this new country," said the brother who had stepped out first.

"Nonsense," said another, "I should be king. I'm the eldest."

Each of the brothers thought he should be king. Soon, a huge fight developed. The brothers threw rocks and clumps of earth at each another. They stamped hard on the ground. Soon, the new island was split in two. Where rocks had fallen, mountains sprang up, and where the brothers had stamped their feet, were lakes. Tiny islands dotted the coast where clumps of earth had fallen into the sea. New Zealand had formed into the islands we know today.

TRUE OR FALSE?

Which of these facts are true and which ones are false?
If you have read this book carefully, you will know the answers.

1. There are several million Philippine monkey-eating eagles alive in the world.

2. Sometimes the head of a giant tortoise is so large that it cannot hide inside its shell.

3. Coconuts from the coco-de-mer tree are the smallest type of seeds in the world.

4. Surtsey Island was formed when a volcano erupted off Iceland.

5. Geysers can be used to generate electricity.

6. Greenland is over three times the size of Texas.

7. Some birds arrive on island shores, perched on drifting bananas.

8. Atlantis may have been an island that was destroyed by a volcano.

9. The people who live on islands in Lake Titicaca build houses and boats from seaweed.

10. In Samoa, islanders eat palolo worms on the beach during the November full moon.

ANSWERS: 1. False 2. True 3. False 4. True 5. True 6. True 7. False 8. True 9. False 10. True

GLOSSARY

● To **adapt** is to change in behavior or form. Animals and plants adapt to fit the conditions they live in.

● An **atoll** is a ring of coral reef that surrounds an underwater island.

● **Conservationists** are people who work to help the environment.

● A **continent** is a very large area of land surrounded by sea. It is too big to be called an island. Africa, Antarctica and Australia are continents.

● **Coral** is a limestone deposit formed by many millions of tiny sea creatures called coral polyps. Some corals look like organ pipes, brains or mushrooms. They can be shades of yellow, pink, purple or green.

● **Endangered** species are animals and plants that are at risk of dying out. Their numbers are so low that there may soon be none left.

● **Environment** is the set of conditions in the area where an animal or plant lives. The plant or animal's survival depends on how well it responds to these conditions.

● To **evolve** is to change over time to suit the surroundings. Plants and animals do this.

● **Extinction** is when the last one of an animal or plant species dies out. This often happens when animals are overhunted by humans or other species, or when they lose their food source.

● **Glaciers** are vast rivers of snow and ice that move slowly down from the peaks of mountains. Many glaciers melted in warmer climates after the Ice Ages and caused world sea levels to rise.

● **Habitats** are the natural surroundings of a plant or animal. Island habitats include coasts, forests, rivers, coral reefs and mountains.

● An **Ice Age** is a period of extreme cold when many glaciers form.

● A **lagoon** is a shallow body of water, usually separated from the sea by a coral reef or sand bar.

● **Lava** is the melted rock that spews out of an erupting volcano. It is red-hot when it first spills out, and turns black as it cools and hardens.

● **Migration** is the moving from one area to another. Many wild animals and birds migrate once a year as the winter grows cold and they look for fresh supplies of food and warmer climates.

● **Natural resources** are the raw materials, like plants or minerals, that make up the necessities and luxuries of life.

● **Pollute** means to poison air, land or water. It is often caused by the waste from industrial activity.

● **Plantations** are tracts of land where crops or trees are planted, often for money.

● **Predators** are animals that hunt, kill, and feed on other animals.

● A **species** is a group of animals or plants with the same characteristics that can breed with each other. Lions are an animal species.

● **Terrain** is the landscape and its special features. Terrain is made up of different kinds of rock, soil, trees, and plants, levels of land and water.

● **Tropical** describes the hot, wet areas on either side of the equator.

● A **volunteer** is someone who offers time and help because they want to.

INDEX

RESOURCES

The Children's Museum of Indianapolis– Coral Reef Adventure,
http://www.cees.iupui.edu/research/ watershed_adventure/
This Web site presents a log written by students who went on a nine-day research mission at the Ten Thousand Islands Refuge in South Florida. The adventure was sponsored in part by The Children's Museum of Indianapolis and The Center for Earth and Environmental Science at IUPUI.

Endangered Island Animals, by Dave Taylor, 1992. The problems faced by endangered island animals such as lemurs, giant tortoises, and land iguanas are explored.

Islands, by Julia Waterlow, 1995. This book examines islands and the threats to island ecosystems.